DEPARTMENT OF THE INTERIOR
FRANKLIN K. LANE, Secretary

UNITED STATES GEOLOGICAL SURVEY
GEORGE OTIS SMITH, Director

Bulletin 699

THE PORCUPINE GOLD PLACER DISTRICT ALASKA

BY

HENRY M. EAKIN

WASHINGTON
GOVERNMENT PRINTING OFFICE
1919

CONTENTS.

	Page.
Introduction	5
Geography	5
Location and extent	5
Topography	6
Climate	7
Forests	7
Water supply	8
Population	8
Communication	8
General geology	9
Character of the rocks	9
Sedimentary rocks	9
Igneous rocks	12
Igneous metamorphism	13
Mineralization	13
Quaternary deposits	15
Economic geology	18
Extent and character of the mineral deposits	18
Gold placers	19
General features	19
Formation	20
Mining methods	21
History of development	23
Summary of mining in 1916	25
Prospecting in adjacent districts	26
Rainy Hollow district	26
Region east of Chilkat Valley	26
Geology	26
Veins and mineralization	26
Iron-ore deposits near Haines	27

ILLUSTRATIONS.

	Page.
PLATE I. Geologic reconnaissance map, Porcupine district, Alaska	6
II. Map of part of southeastern Alaska showing location of Porcupine district	8
III. *A*, Glaciated mountains at the head of McKinley Creek, McKinley Creek canyon in middle ground; *B*, Spruce and hemlock forest near Pleasant Camp	10
IV. *A*, Klehini River at evening high water, a few miles above Pleasant Camp; *B*, Thin-bedded limestones and slates, Porcupine Creek	11
V. *A*, Upper end of abondoned canyon of McKinley Creek; *B*, Lower end of same abandoned canyon showing glacial till exposed by placer workings	20
VI. *A*, Cable tramway formerly used in stacking boulders, Porcupine Creek; *B*, Crane hoist stacking boulders behind cribbing at sides of sluice box on McKinley Creek bench placer	21
VII. *A*, Diverting flume, Porcupine Creek; *B*, Worked-out section of modern McKinley Creek canyon	22
VIII. *A*, Flume and spillway, Glacier Creek dam; *B*, Detail of Glacier Creek flume, showing use of riffle blocks	23

4

THE PORCUPINE GOLD PLACER DISTRICT, ALASKA.

By Henry M. Eakin.

INTRODUCTION.

The Porcupine gold placer district includes the drainage basins of the westerly headwater tributaries of Chilkat River, which enters Chilkat Inlet, an arm of Lynn Canal, near Haines, a small town about 75 miles northwest of Juneau. Gold was first discovered in this area in 1898. Productive mining was begun the following year and has been continued to the present time with such success that the district has ranked with the most important placer fields of southeastern Alaska.

The region was first visited by a geologist of the United States Geological Survey in 1899, when an exploratory expedition on the way from Pyramid River to Eagle followed the old Dalton trail up Chilkat River and past the then newly discovered placers on Porcupine Creek.[1] In 1903 C. W. Wright made a more extended study of the region and outlined the principal features of its geology and mineral deposits.[2]

The writer visited the district early in the summer of 1916, spending two weeks in the examination of properties then under development and in the study of bedrock and glacial geology related to the formation of the placer deposits.

The writer is indebted to the residents of the district, who extended generous hospitality during his stay and facilitated his work in every possible way. Acknowledgment is also due the earlier investigators, whose published papers have been drawn upon in assembling the data of this report.

GEOGRAPHY.

LOCATION AND EXTENT.

The general position of the Porcupine placer district in relation to the geographic features of southeastern Alaska is shown on Plate II.

[1] Brooks, A. H., A reconnaissance from Pyramid Harbor to Eagle City, Alaska: U. S. Geol. Survey Twenty-first Ann. Rept., pt. 2, pp. 374–376, 1900.

[2] Wright, C. W., The Porcupine placer district, Alaska: U. S. Geol. Survey Bull. 236, 1904.

The accompanying geologic reconnaissance map (Pl. I) shows this district and some additional territory, covering an area about 25 miles square which centers near latitude 59° 25' N. and longitude 136° 10' W. The principal belt of mineralization that has produced placer deposits runs northwestward across the area shown on the map a little below its center. A secondary area of mineralization occurs in the vicinity of Bear Creek in the northeastern part of the region mapped.

TOPOGRAPHY.

The district is drained by four westerly headwater tributaries of Chilkat River, which flows into an arm of Lynn Canal, 40 miles southeast of the main placer area. These streams, named in order from north to south, are Bear Creek and Klehini, Salmon, and Takhin rivers. Of these the Klehini is much the largest, and its basin contains the most productive placers. All these streams flow through broad, glaciated valleys floored with thick deposits of gravel and boulders. Their grades are steep, and their swift-flowing waters are generally divided among numerous shallow shifting channels. These features are due to the presence of glaciers of considerable size on most of their headwaters. The topography of the district is typical of the coastal mountain province of southeastern Alaska in its strong relief and glacially developed features. (See Pl. III, A.) Summit elevations along the principal divides generally range between 3,000 and 5,000 feet, but individual peaks rise 6,000 to 7,500 feet above sea level. The topography has a smoothed aspect up to an altitude of about 3,000 feet, marking the level reached by ice during earlier periods of glaciation. Above this level more rugged forms abound, culminating in the serrate ridges and peaks of the high divides. The valley heads throughout the district have the form of glacial cirques. These occur at great elevations in the main divides, and the valleys leading from them do not descend with an even slope but in a series of steps, causing an alternation of falls or cascades and relatively sluggish reaches in the streams that drain them. The amount of glacial scour has varied with the size of the valleys, and the bottoms of main valleys are generally much below those of their tributaries. Streams draining these hanging valleys have generally intrenched themselves, producing an interesting series of canyons, and have built alluvial fans out into the valleys which they join.

Features formed directly by deposition from the ice are almost entirely absent except in the immediate vicinity of the present glaciers. There is a little morainic topography along the north side of the upper Klehini Valley, and there are some high terraces along the south side of the same valley that may in part be lateral moraines.

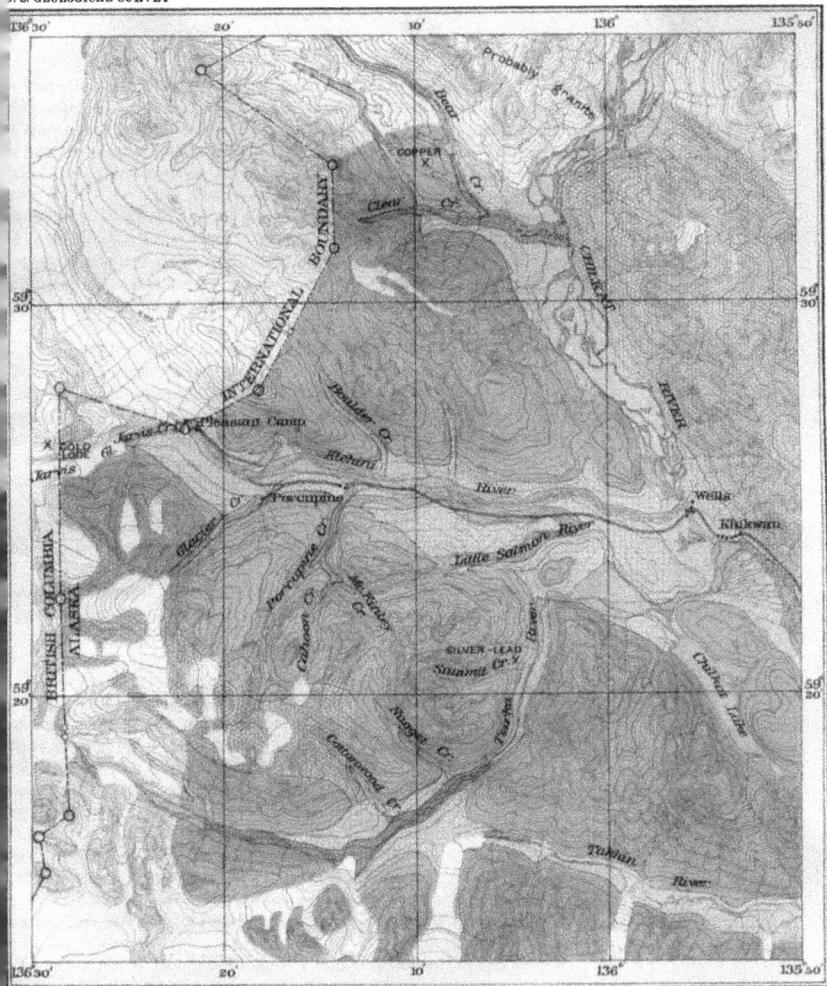

EXPLANATION

Stream gravels

Bench gravels
including glacial deposits

Auriferous gravels

Gold placers

Diorite

Shale and limestone
(*Carboniferous*)

Metalliferous prospect

Base from map by International
Boundary Commission

Scale 1:250,000

Geology by C. W. Wright, 1904
and H. M. Eakin, 1916

GEOLOGIC RECONNAISSANCE MAP OF PORCUPINE DISTRICT, ALASKA

CLIMATE.

The climate of the Porcupine district is intermediate in general character between that of the excessively humid, equable climate of the coastal provinces, where the full influence of the Japan current is felt, and that of the interior of Alaska, where relative aridity and great seasonal variations in temperature prevail. The winters are long and fairly cold. Heavy snows may occur as early as October and as late as April, and its accumulation may reach a depth of 10 or 15 feet in the valleys and much more in the uplands. The line of perpetual snow is about 5,000 feet above the sea, but snow banks in protected situations at much lower elevations may persist for the greater part of the summer. The summers are shorter than the winters, but they are for the most part warm and sunny. As the region is sheltered behind the great ice fields of the St. Elias Range the southerly and westerly winds that bring the summer rains to most of southeastern Alaska give fair weather in the Porcupine region. The weather in the valleys during the writer's visit in the later part of June was frequently uncomfortably hot. The season available for placer operations is much longer than in interior Alaska, as it extends from early May to late October.

The climate of the district favors agriculture to an unusual degree in comparison with other sections of southeastern Alaska. Where the soil is favorable at low elevations practically all the common vegetables, small fruits, and cereals are grown successfully, and the gardens of Haines and the Chilkat Valley are justly celebrated for the abundance and fine quality of their strawberries.

FORESTS.

The slopes of the region up to an altitude of 2,000 or 2,500 feet are generally covered with a dense forest, mainly of spruce and hemlock but with a few cedars and pines in places. The valley bottoms near the streams support an abundant growth of cottonwood which intermingles with the conifers toward the bordering slopes. The trees are generally of large size, suitable for lumber, and the supply far exceeds any possible demand of the local mining industry. (See Pl. III, B.) Large tracts of timber in the Klehini Valley in Canadian territory are held for future development for lumbering.

The hemlock is especially adapted for piles and also makes a good grade of lumber. The spruce meets with favor for flume and sluice-box lumber, and the cottonwood is preferred for riffle blocks owing to its tough fiber.

In most places the forests have a dense undergrowth of alder, "devil's-club," blueberry, fern, and other plants, which, by concealing the bedrock and making walking difficult, hinders geologic study and bedrock prospecting.

WATER SUPPLY.

The snow fields and glaciers of the high divides furnish an abundant water supply in all the streams throughout the summer. In fact, the chief difficulty encountered in mining has been that of handling the copious discharge of the glacial streams during the warm summer days, and the repeated failures of diverting flumes has detracted much from the profit of operations. The strong relief of the region and the steep grades of the streams permit water power and hydraulic systems to be developed at an unusually small outlay.

The streams are generally low during May and June, October and November, and the high-water season intervenes between these spring and fall periods. In winter the streams freeze over and are covered with snow, but flowage beneath the ice is continuous. In summer there is a great daily variation in discharge in the glacial streams, with low water in the morning and high water in the evening. (See Pl. IV, A.)

POPULATION.

The population of the district is made up almost exclusively of the forces engaged in placer mining. Therefore it varies greatly in number with the seasons and from year to year. During the summer of 1916 an average of about 60 persons were within the district. The principal settlement is Porcupine, near the mouth of Porcupine Creek. At the international boundary on Klehini River near the mouth of Jarvis Creek is the abandoned post of the Northwest Mounted Police, known as Pleasant Camp. In winter, when mining operations are at a standstill, almost everyone leaves for the settlements on the lower Chilkat or elsewhere. The principal permanent settlements of the region are Klukwan, a native village near the mouth of Klehini River, and Haines, an important town at the base of Chilkat Peninsula. In addition there are a score or more of permanent homes on the ranches of Chilkat Valley between Haines and Klukwan.

COMMUNICATION.

Haines is on an arm of Lynn Canal and is a port of call for many of the boats plying the inside passage to and from Skagway. Mail service is thus maintained throughout the year. The service of the Government cable also is available at Haines. From Haines to Klukwan and thence to Porcupine there is a Government wagon road that is generally open for traffic. In 1916 it was available for automobiles as far as Klukwan. Beyond this point it runs partly over lowlands and along gravel bars of Klehini River that are subject to overflow at high stages of the river. However, horse-drawn vehicles use it

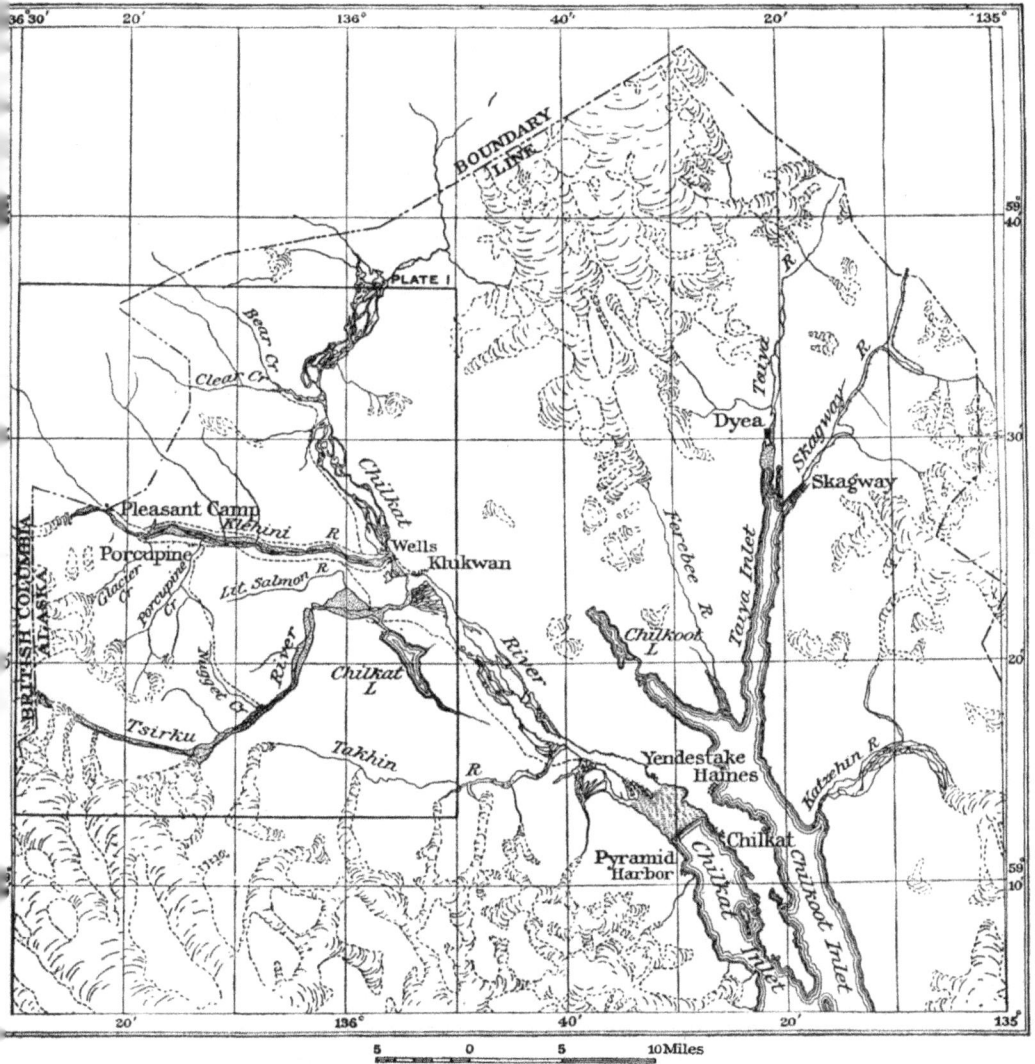

MAP OF PART OF SOUTHEASTERN ALASKA, SHOWING LOCATION OF PORCUPINE DISTRICT.

throughout the summer, regardless of the height of the water. The road has played an important part in the development of the district by permitting the haulage of the heavy equipment of the hydraulic plants and by lowering the cost of supplies.

GENERAL GEOLOGY.

CHARACTER OF THE ROCKS.

The solid rocks of the district consist mainly of altered sedimentary types and include limestones, slates, phyllites, and coarse clastic rocks. Silicification of the limestone has produced quartzitic rocks locally, and in a few places excessive alteration of the clastic rocks has given them the character of schists.

The general area of sedimentary rocks is bordered on the northeast by the great diorite belt of the Coast Range, and outlying masses of diorite are intruded into the sedimentary rocks for some distance southwest of the principal contact. Small diabase dikes also intrude the bedded rocks locally. A narrow band of greenstone schist that is probably of igneous origin, associated with black slates, was noted by Wright[1] at a single locality on the north bank of Klehini River.

The general distribution of the sedimentary rocks and dioritic intrusives is shown on the accompanying geologic reconnaissance map (Pl. I), which is reproduced with minor corrections from a map drawn by Wright.[2]

Unconsolidated deposits mantle much of the region below an altitude of 3,000 feet. They include gravels laid down by the streams in the present cycle of erosion, outwash gravels above the present stream levels, and morainic materials that are distributed in erratic fashion up to the highest levels reached by glaciers during their maximum development. Only the general distribution of present stream gravels and terrace deposits is indicated on the map (Pl. II).

SEDIMENTARY ROCKS.

The sedimentary rocks are mainly limestones and slates and their metamorphic equivalents. Thin sandstone beds and fine conglomerates occur in the slates at one locality.

The limestones and slates are in places clearly interbedded and of like age. Elsewhere limestone beds occupy considerable areas without associated slates, and some areas of slate are devoid of limestone beds. Massive limestone beds occur in a broad band along the south

[1] Wright, C. W., The Porcupine placer district, Alaska: U. S. Geol. Survey Bull. 236, p. 17, 1904.
[2] Idem, pl. 5.

margin of the Klehini Valley, forming the walls of Porcupine and Glacier Creek canyons at their lower extremities. In line with this band a large area in the mountains north of Jarvis Glacier is occupied by similar beds. For several miles up the Klehini Valley above Jarvis Creek only limestone outcrops were found, although in places sections several hundred feet thick are exposed. These rocks, where least altered, are light gray in color, are locally fossiliferous, and show a fairly regular thick-bedded structure. In places, as just east of Porcupine Creek and where the road going eastward from Porcupine leaves the Klehini bottom lands, silicification has occurred, producing quartzitic rocks that resemble the limestones in texture and color. On Glacier Creek and in the Klehini Valley the limestones are largely recrystallized, producing a light-colored compact marble.

Where limestone and slate are interbedded, as on Porcupine Creek above the marble canyon, the whole series is dark and both limestone and slate have a thin-bedded, platy structure (Pl. IV, B). A graduation in composition is evident, the limestones grading into calcareous slates and these into more purely siliceous beds. Graphitic material is present in all the beds, and in some of the slates it is very abundant, giving them a coal-black color and a soft, friable texture.

Thin beds of limestone and calcareous slates were noted also on the upper Porcupine, near the mouth of McKinley Creek, and in the road a few miles east of Porcupine. It would appear that there is among the sedimentary groups an important assemblage of slates distinguished by calcareous phases, among which are some nearly pure limestones.

Slates without interbedded limestones are exposed on McKinley and Cahoon creeks, in the upper part of Glacier Creek valley, and in Jarvis Creek valley below the glaciers. In each locality the rocks are thin-bedded, dark, clastic sediments. The most obvious distinction between different beds is in their graphitic content. In this respect they grade from a soft, richly graphitic type to hard, relatively pure quartzose slates.

A group of coarse-grained sedimentary rocks, including thin-bedded sandstones and slates, and a fine-grained conglomerate bed are exposed on the south slope of Jarvis Valley near the international boundary at an altitude of about 2,000 feet. Lower down on the same slope limestones of the massive type crop out. The clastic beds of this locality were not encountered elsewhere in the district, and their relation to the other terranes is not definitely known. The massive limestones there, locally at least, dip southwest. If this indicates a general structure the conglomerate and associated rocks overlie them.

A. GLACIATED MOUNTAINS AT THE HEAD OF McKINLEY CREEK

McKinley Creek Canyon in middle ground.

B. SPRUCE AND HEMLOCK FOREST NEAR PLEASANT CAMP.

A. KLEHINI RIVER AT EVENING HIGH WATER A FEW MILES ABOVE PLEASANT CAMP

The stream was fordable here earlier in the day.

B. THIN-BEDDED LIMESTONES AND SLATES, PORCUPINE CREEK.

The structure of the sedimentary rocks is apparently much more complex and irregular than it is generally in southeastern Alaska. The beds are steeply tilted wherever observed, dips of 75° to 90° being common. The apparent trend of the great limestone band that crosses the lower courses of Porcupine and Glacier creeks is northwesterly. The limestones of Klehini Valley above Jarvis Creek also strike northwest, and this is the direction of elongation of the larger igneous bodies of the region. However, in many places in Klehini and Jarvis valleys where observations on the structure of the slates were taken the strike is northeast. The apparent differences in structure of the slate and massive limestone series adjacent to Klehini River suggest that the limestone may be a younger formation that overlies the slates unconformably. This interpretation is supported by the fact that the slates have strong compressional structures, whereas near-by limestone beds are massive and unsheared. However, the stratigraphy and structure of the whole assemblage are so complex and obscure that much additional work must be done before the relations and sequence of the various terranes will be revealed.

The general age of the bedded rocks of the region is indicated by fossils found in the limestones on lower Porcupine Creek.

A small collection made by Wright[1] in 1903 was doubtfully identified as lower Carboniferous (Mississippian) by G. H. Girty, an identification which was corrected the following year when larger collections were made at Pybus and Herring bays.[2] The first collection, which was obtained on Porcupine Creek, contains the following species with revised identifications:

> Crinoid fragments.
> Productus aff. P. mammatus.
> Productus aff. P. gruenwaldti.
> Spirifer aff. S. marcoui and S. musakheylensis.
> Camarophoria aff. C. margaritovi.

The same fauna is more completely shown in collections from Saginaw Bay, Kuiu Island, and the facies appears to be that of the *Spirifer arcticus* zone, believed to correlate with the Russian Artinskian (of late Pennsylvanian or early Permian age).

As already stated, there are structural grounds for considering the limestones on lower Porcupine Creek to be the youngest of the stratified rocks of the region, except, perhaps, the conglomerate and associated finer clastic beds noted at the locality south of Klehini River, near the international boundary. Though the limestone and asso-

---

[1] Wright, C. W., op. cit., p. 16.
[2] Wright, C. W., A reconnaissance of Admiralty Island: U. S. Geol. Survey Bull. 287, p. 143, 1906.

ciated beds may represent the upper Carboniferous and earlier Paleozoic, the coarser elastic series may correspond with the Mesozoic beds of Admiralty Island, which they resemble in a general way.

The correlation of the rocks of the Porcupine district with terranes that occur on Admiralty Island emphasizes a feature of the Coast Range intrusions noted by Spencer [1]—namely, the edging over of the igneous mass toward the north, whereby one after another of the formations present in the Juneau gold belt is terminated.

The last of these formations to be cut out by the diagonal boundary of the diorite is a group of basic volcanic rocks that make up the Chilkat Peninsula and extend up the east side of Chilkat Valley halfway to Klukwan. They have the general characteristics of the group of metabasalts that lie next to the diorite rocks in the Berners Bay region, which Knopf [2] assigns to the Jurassic or Lower Cretaceous.

The contact between these rocks and the sedimentary rocks that extend into the Porcupine district is covered by the alluvium of the Chilkat Valley. Their difference in physical alteration points to an unconformable relation and is entirely compatible with the age assignments that have been indicated.

The rocks of the Porcupine district are therefore not to be correlated with those of the Juneau gold belt. The only point of similarity in the two regions is their proximity to the intrusive rocks of the Coast Range. It seems clearly apparent that the mineralization in both regions is due to the influence of the intrusive rocks and not to any inherent qualities in the bedded rocks which were invaded.

IGNEOUS ROCKS.

The igneous rocks of the district are apparently all intrusive. The bodies large enough to be represented on the geologic map are diorites connected with the deep-seated intrusive complex of the Coast Range. In addition there are a few small diabase dikes whose age and relation to the other intrusive rocks have not been determined.

The diorite of the northeastern part of the area is part of the main Coast Range belt, which extends uninterruptedly from lower British Columbia northwestward past this region and across the Alsek drainage basin. A small outlying band of diorite forms the crest of the range north of Jarvis Glacier; a larger one extends northwestward from upper Salmon River, across the heads of Porcupine and Glacier creeks and beyond the international boundary.

---

[1] Spencer, A. C., The Juneau gold belt: U. S. Geol. Survey Bull. 287, p. 12, 1906.
[2] Knopf, Adolph, Geology of the Berners Bay region, Alaska: U. S. Geol. Survey Bull. 446, p. 19, 1911.

The diorites of the district vary considerably in texture and composition, but the most prevalent types are moderately coarse, equigranular light-gray rocks composed dominantly of plagioclase feldspar and hornblende with smaller amounts of biotite and quartz and with accessory apatite, titanite, and magnetite. The rocks of the main Coast Range belt are chiefly of this character, but local facies show either quartz or mica or both in unusual abundance, giving them the character of quartz diorites and quartz-mica diorites.

The narrow band at the head of Porcupine Creek is described by Wright,[1] as follows:

The narrow belt at the head of Porcupine Creek contains more biotite, and segregations of hornblende and mica are often prominent, causing large dark spots, which are locally characteristic. The continuation of this belt to the east at Cottonwood Creek is characterized by a microcline feldspar and a larger amount of quartz. Under the microscope some of the minerals are seen to have been crushed, indicating that since the intrusion of the diorite they have been subjected to pressure and movement. This diorite rock is locally termed granite, which it resembles very closely and from which it can be distinguished only by careful examination.

IGNEOUS METAMORPHISM.

The metamorphic effect of the diorites upon the inclosing rocks is generally limited to a zone only a few hundred feet wide along the contact. As pointed out by Wright, the slates have been " baked and altered to a flinty hornstone " in such metamorphic zones, and the limestones have been recrystallized, forming white, compact marble. A more widespread phenomenon, also due to the influence of the invading magmas, is evident in the mineralization of the district and adjacent areas.

MINERALIZATION.

The very general mineralized condition of the sedimentary rocks of the region is well described by Wright,[2] who says:

The sedimentary rocks have all been more or less mineralized by stringers and veins of quartz and calcite, but an especially noteworthy impregnation of iron sulphides forms an interrupted zone of mineralization in the southern portion of the sedimentary series. The sulphides in the slates occur as films or frequently as lenticular masses a few inches in width, parallel with the bedding. Two samples of the mineralized slates—one an average across several feet and the other from a rich seam—gave, respectively, $0.41 and $2.48 per ton in gold. Samples from near the mouth of the Porcupine, where the slates are apparently unmineralized, taken by Mr. Brooks during his short visit to this region in 1899, gave traces of both gold and silver.

---

[1] Wright, C. W., The Porcupine placer district, Alaska: U. S. Geol. Survey Bull. 236, p. 17, 1904.
[2] Idem, pp. 17–18.

The quartz veins are not very abundant, and as a rule are short and small, often merely stringers parallel with the structure of the slates. A few which cut directly across the formation carry galena and sphalerite, with a small amount of chalcopyrite, and, though quite narrow, often persist for considerable distances. Calcite veins, which are more numerous than those of quartz, are usually a foot or more in width and are often weathered to a light-brown color on the surface, while of a bluish color and fine granular structure when freshly broken. They often carry cubes of pyrite, which occasionally measure an inch across. From veins of this nature up McKinley Creek some native gold has been reported.

Besides the small veins a quartz ledge 100 feet wide outcrops at an elevation of 2,000 feet on the ridge south of Porcupine. Although apparently quite barren, a small sample from this gave an assay value of $5.28 in gold. A similar ledge occurs across the Klehini at 1,500 feet elevation, on the ridge west of Boulder Creek. About 2 miles below Porcupine is a third mineralized deposit rich in sulphides, with calcite as gangue mineral, but a sample taken here gave an assay value of only 41 cents.

The general zone of mineralization from which the placers of Klehini and Salmon river basins have been derived appears to be elongated in a northwesterly direction, extending from a point south of Salmon River across the basins of Porcupine, Glacier, and Jarvis creeks and into the mountain mass north of Jarvis Glacier. The richness of mineralization varies from place to place, and there are large areas of lean or barren rock in this zone. But there are also remarkably large areas in which sulphide minerals are generally abundant and which are reported to yield substantial returns in gold on assay. A band of this sort that cuts across Cahoon Creek near its mouth in the slate formation shows abundant quartz veining and sulphide mineralization for a width of nearly 1,200 feet. Random samples taken across this belt are reported to assay from a trace up to several dollars a ton in gold, a large number of assays going between $1 and $2 a ton.

The point farthest northwest along the general zone of mineralization at which gold has been found is in the mountains north of the lower end of Jarvis Glacier, west of the international boundary. Here, at an elevation of about 4,500 feet are several gold quartz ledges in granitic country rock about 300 feet above its contact with limestones. The lowest and apparently the richest of the ledges is traceable for over 2,000 feet along the strike. It ranges from 1 to 4 feet in width. Assays of numerous samples taken across the full width of the vein are reported to give from a few dollars to $70 a ton; the higher tenor is generally found where the vein is relatively narrow. A silver content of a few ounces to the ton is indicated in most samples. Picked specimens give much higher assay values.

Southeast of the main placer area, on the slope north of Salmon River, are a number of narrow silver-lead veins that represent still another type of mineralization in the district. So far as known the largest are less than a foot in width. The maximum metal content of the veins is said to be about $3 a ton in gold, about 60 ounces a ton in silver, and about 35 per cent of lead. One sample shows a copper content of nearly 3 per cent.

The mineralization that has furnished the placer gold in the vicinity of Bear Creek has not been studied geologically. Prospectors report that a zone along the ridge west of Bear Creek is heavily mineralized and that in places it contains commercial grades of copper ore. A specimen from this deposit, taken apparently from a vein a few inches wide, contains pyrite, pyrrhotite, chalcopyrite, and sphalerite, together with a little quartz gangue. The zinc sulphide occurs as a marginal band, and the main mass of the vein is dominantly chalcopyrite and pyrrhotite. This type of mineralization seems to be allied with that of the Rainy Hollow district, at the head of Klehini River, in Canadian territory, which is briefly described elsewhere (p. 26).

QUATERNARY DEPOSITS.

The Quaternary deposits of the region include both glacial and fluviatile types. Deposits made directly by ice are of surprisingly rare occurrence in a region that has been so extensively glaciated. Moraines are found about the ends of the present glaciers but are almost wholly lacking elsewhere. Glacial till is more widespread, and small deposits of this character occur here and there wherever conditions favored lodgment and protection from subsequent erosion. On some of the glaciated slopes till deposits form a thin veneer; on others only a few scattered erratic boulders are to be found. Till is more abundant in or near the bottoms of the valleys. It enters largely into the composition of terraces along the upper Klehini River, which have a broad extent near the international boundary. A considerable area at the north side of the valley opposite Glacier Creek behind an outlying bedrock ridge is underlain by till, the surface of which in places is marked by kettles and other morainic features. Abandoned canyons in the valleys of Glacier and Porcupine creeks are filled with till.

The fluviatile deposits include the gravels of the present valley bottoms, alluvial fans of tributary streams, and older gravels that form extensive terraces along the lower courses of the larger streams. The trunk streams are at present aggrading their valleys. Their load, which is derived mainly from glacial sources, includes heavy boulders, gravels, sands, and silts. The coarse part is deposited mainly along a broad zone in each valley, which is traversed and reworked by the numerous shifting channels of the anastomosing streams. Such zones are clear of vegetation and at low stages expose a great expanse of gravel and boulder bars, shaped into fantastic forms by the transitory channels of the last flood and varying in texture from place to place according to strength of the currents that brought their materials into place.

The higher floods are not confined to the channels of the barren zones of the valleys but overflow the timbered tracts of bottom lands along the valley sides. The overflow waters carry only the finer grades of detritus, so that though the main flood way is built up with boulders and gravels other areas of the bottom lands are occupied only by sands and silts. The barren flood ways shift their positions from side to side, so that boulder deposits may come to rest on top of silt beds and vice versa.

The heaviest boulders find permanent lodgment near their glacial sources. The smaller boulders and gravels and much of the sands and silts are carried farther downstream, the distance depending upon their fineness. Thus there is a gradation from coarse to fine materials in contemporaneous deposits along the main flood way from source to mouth. The delta of the Chilkat is being built up entirely of very fine gravels, sands, and silts.

Much of the valley of Chilkat River and perhaps of its main tributaries was scoured by ice to a considerable depth below sea level. As the glaciers retreated, so long as they continued to end in the sea their detritus must have been deposited with little or no assortment on the estuary floors. When their retreat brought their ends up to sea level fluviatile deposition and delta building began. Further retreat of the glaciers and the extension of deltas brought the processes of assortment into play. As the texture of the débris deposited at any point along the stream is a function of the distance to its glacial source the vertical section of the deposits, where considerable aggradation has occurred during the ice retreat, shows progressively finer materials toward the surface.

This outline of the processes that have operated in the aggradation of main valleys of the region indicates the probable complexity of the valley fillings generally, below the present flood levels. The logs of drill holes in the Klehini Valley show that this probable complexity is fully developed in some places at least.

The alluvial fans of tributary streams lie at higher elevations than the adjacent deposits of the trunk streams, and therefore they have been interpreted by some as terrace deposits laid down in adjustment to a higher base-level of erosion than the present. They are not level topped, as terrace deposits should be, but slope uniformly away from the points where the tributaries leave the confines of their own hanging valleys and debouch upon the graded plains of trunk streams.

The fans are marked by numerous shallow channels leading from summit to margin, one or more of which convey the present drainage. The fans have been built up by the present streams out of materials such as they now carry and in adjustment to the main streams in the present cycle of erosion.

The appearance of a level terrace has been developed between the alluvial fans of Glacier and Porcupine creeks where the town of Porcupine stands. These creeks are so near each other that their fans come together some distance out from the edge of Klehini Valley. The rudely triangular area bordered by the two fans and the south slope of the main valley have been filled level with silt and peaty material.

Alluvial deposits that are unrelated to the present cycle of erosion form terraces at several places in the district and the adjacent regions. The east side of the Klehini Valley for several miles above Jarvis Creek has broad terraces, from 100 to 300 feet above the present stream. At the apex of the alluvial fan of Porcupine Creek there are high gravels that appear to be the remnants of an old fluviatile terrace, and a similar feature occurs at the mouth of Boulder Creek. The broad plateau that lies between Klehini and Little Salmon rivers is apparently underlain by alluvium.

The origin of these features has not been fully determined. They may have been built as the result of retreat of tributary glaciers before the trunk glacier had receded past their junctions. According to this hypothesis the tributary streams filled in the lower ends of their valleys against the side of the main glacier, and when the ice had retreated sufficiently they intrenched themselves in the deposits which they had just made and which have persisted as terraces up to the present time. In some places the glacial evidence seems to accord with this hypothesis, but it is possible that the region has been uplifted in late Quaternary time and that these deposits were laid down nearer sea level than their present altitude would suggest. Some recent elevation of the land about Lynn Canal and Chilkat Inlet is clearly indicated by marine terraces, but the extent of uplift directly evident is far short of that required by this hypothesis. Farther south, in the vicinity of Juneau, sea shells and old beach deposits occur up to elevations of 600 feet or more above the present sea level.[1] More extensive uplift is noted along the coast southwest of the district where, near Yakataga, Pleistocene marine deposits occur up to a known elevation of over 5,000 feet above sea level.[2]

More complete data regarding the structure of the high terrace deposits and the extent of recent uplift in the Porcupine region itself will be required before the origin of these deposits can be definitely stated.

---

[1] Spencer, A. C., field notes, 1916.
[2] Maddren, A. G., Mineral deposits of the Yakataga district, Alaska: U. S. Geol. Survey Bull. 592, pp. 131–132, 1914.

ECONOMIC GEOLOGY.

EXTENT AND CHARACTER OF THE MINERAL DEPOSITS.

Gold is widely distributed in bedrock in the Porcupine district, for it occurs in placer form in the gravels of Bear Creek and Klehini, Salmon, and Takhin rivers, as indicated on the map (Pl. II). Workable placers, however, have been found only in the Klehini and Salmon river basins, on tributary streams that have had a peculiar erosional history, as described elsewhere (p. 19).

The commercial possibilities of gold deposits in bedrock have been investigated only to a very slight extent. Erratic prospecting in the vicinity of the richer placers has shown that considerable bodies of rock are gold bearing, and samples have been assayed which show a gold content of $1 to several dollars a ton. Although the existence of ore bodies of such size and richness as to be at present available for development has not yet been demonstrated, the showing that has been made should encourage further search for commercial lode deposits. The type of deposit to be looked for is that of large low-grade stockworks and mineralized zones in the slates. Quartz veins and abundant pyrite in the rocks should prove a rude indication of the presence of gold.

Gold quartz lodes of more compact form than the type suggested above have been found just west of the international boundary in the mountains north of Jarvis Glacier, at an altitude of about 4,500 feet. One of these lodes consists of a quartz vein that is 1 to 4 feet wide and is traceable for about 2,000 feet. The compactness of the vein is due to the fact that it cuts a diorite country rock which, on being deformed, develops a few large widely spaced fissures rather than numerous smaller ones, as do the slates. The location of the lode that is best developed is indicated on the map, although it is in Canadian territory. It is mentioned in this report because of its evident relationship to the deposits on the American side of the boundary.

Silver, lead, and copper occur in the district, but as yet they show little promise of commercial importance. Silver and lead occur together in some narrow veins that have been slightly developed on the north side of the Salmon River valley near Summit Creek. A little copper also is found at this locality. These veins are rich in the metals but are too small for profitable exploitation.

Copper deposits are reported to occur in the area between Bear and Clear creeks in the northeastern section of the district. Specimens from this locality show iron, copper, and zinc sulphides in vein form. The extent and richness of the deposits have not been determined.

GOLD PLACERS.

GENERAL FEATURES.

The distribution of auriferous gravels and gold placers is indicated on the map (Pl. II). The bedrock source of the gold is discussed under "Mineralization" (pp. 13–15), and the history of its concentration in placers is discussed on pages 20 and 21. Peculiarities of form, distribution, and topographic relations of the placers are here described and explained.

The only placers regarded as workable at present are those of the Glacier and Porcupine drainage basins. The placers of Nugget Creek were formerly worked in a small way, but this ground is now abandoned. Extensive tracts of auriferous gravels were formerly held on Bear Creek, on the Klehini bottoms near Porcupine, and in the Salmon River valley near Nugget Creek. Extensive prospecting in these areas, by manual methods on Bear Creek and in the Salmon River valley and by drilling on the Klehini bottoms, failed to develop workable deposits, and the claims were abandoned.

The gold of the placers is generally well worn, of a bright color, and of medium fineness, assaying on the average about $17 an ounce. The texture of the gold dust shows a lack of assortment, the particles ranging in size from flour gold to nuggets weighing several ounces. Flour gold is present in unusual amount along with the coarser particles, and it occurs also in deposits of glacial mud or rock flour associated with the placers. This distribution is believed to be due to glacial abrasion during certain epochs in the erosion of the valleys and concentration of the placers. Great quantities of mineralized rock have been scoured out by the action of the ice. The rock removed by this process was largely in the form of rock flour; the gold which it contained must likewise have been largely reduced to a flour-like powder. The load of the streams that concentrated the placers has at all times been composed dominantly of glacial detritus. Though perhaps the most of the flour gold has been carried beyond the limits of the placers, much has found lodgment with the coarse gold and in the fine sediments that have been deposited here and there in the placer-bearing valleys.

A sample of glacial mud taken at random on Porcupine Creek near the lower end of its canyon gave an assay of 1.10 ounces of gold a ton.[1] Other assays of sediments reported by local operators show that considerable gold is commonly present in them.

The flour gold is practically all lost in the present mining operations. As the placers are all of similar genesis, great importance should be attached to further investigation of the distribution of

---

[1] Wright, C. W., op. cit., p. 20.

flour gold and to the adaptation of methods to its recovery, if possible, along with the product now saved.

FORMATION.

Elements in the history of erosion have already been presented (p. 19) that show the main valley fillings and high terraces to have been built up by processes that did not favor the concentration of placer gold. The placers all lie in the smaller tributary valleys. The history of this erosion includes the origin of the placers and explains their character and distribution.

During the periods of more extensive glaciation of the region, both trunk and tributary valleys were deeply scoured by the action of the ice, but the main valleys were generally lowered below the level of their tributaries, giving to the tributary valleys the character of hanging valleys. When the ice had retreated the streams from the hanging valleys found abrupt declivities in their courses at the margins of the main valleys, which brought about the erosion of canyons in tributary valleys and the deposition of detritus in the form of alluvial fans along the margins of the main valleys.

In places in the Porcupine Valley at the side of the present canyon there are so-called bench deposits, which consist of stream gravels overlain by glacial detritus. It is evident that these deposits occupy sections of a canyon, older than the present but of similar origin, which was in some places followed and in others missed by the course of the stream when the last intrenchment began. (See Pl. V, *A* and *B*.) Two distinct ice advances are thus indicated, each of which was followed by intrenchment of the hanging-valley streams.

The modern canyon has been eroded to a lower level than the earlier throughout the middle and upper sections of Porcupine Valley. Near the lower end of the valley the conditions are not so simple, and it seems likely that the stream accomplished considerable intrenchment along more than two courses. The identity and relations of the different bedrock canyons at this place could not be fully deciphered from the available exposures. It is clear, however, that the Porcupine was controlled by a lower base-level than that afforded by the present position of Klehini River at the time of maximum intrenchment in each position.

In the Glacier Creek valley also there is evidence of two distinct ice advances with a period of stream erosion intervening. An extremely deep and narrow bedrock gorge filled with glacial detritus has been traced for some distance along the lower part of the valley beneath the modern stream gravels. A base-level of erosion much lower than the present is indicated the same as in the lower Porcupine Valley. The upstream extension of this older canyon

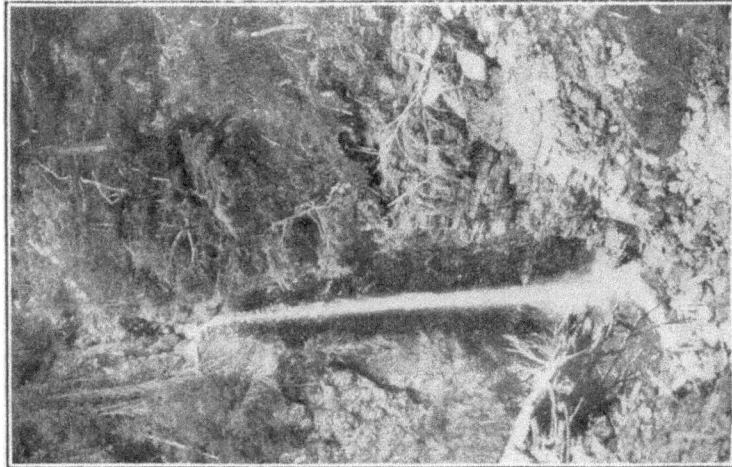

A. UPPER END OF ABANDONED CANYON OF McKINLEY CREEK.

The glacial filling has been sluiced out for a short distance. The water of the falls comes from a hydraulic plant and spills over the rim of the modern canyon into McKinley Creek. (See also Pl. VII, *B*.)

B. LOWER END OF SAME ABANDONED CANYON SHOWING GLACIAL TILL EXPOSED BY PLACER WORKINGS.

BULLETIN 699 PLATE VI

A. CABLE TRAMWAY FORMERLY USED IN STACKING BOULDERS, PORCUPINE CREEK.

B. CRANE HOIST STACKING BOULDERS BEHIND CRIBBING AT SIDES OF SLUICE BOX ON McKINLEY CREEK BENCH PLACER.

is not now evident. It may have been destroyed by later ice erosion in the Glacier Creek valley, or it may have been followed by the stream in its last intrenchment. The narrowness of the valley favors the latter interpretation.

The concentration of placer gold took place in conjunction with the intrenchment of the streams in the hanging valleys wherever their courses traversed zones of mineralized bedrock. The placers are therefore of two distinct ages, corresponding with the separate periods of stream erosion.

The concentrations are generally found in a thin stratum of stream gravels lying on the bedrock bottoms of the canyons. Locally, as below the falls on McKinley Creek, gold has been found on bare bedrock practically without associated gravels. The gold-bearing stratum is generally overlain by barren or very low grade gravels that are progressively deeper downstream. Their deposition is due to the rising base-level afforded by the aggrading Klehini River.

The stream gravels in the bottoms of the older canyons, which are gold bearing in places, are overlain by glacial detritus the depth of which is generally equal to the height of the canyon walls. In the lower section of Glacier Creek modern gravels overlie the glacial fill of the older canyon, which extends to a great depth below the present stream and has a much steeper grade.

At the lower ends of the modern canyons, where the alluvial fans begin, there are certain concentrations that extend out into the gravel deposits somewhat above bedrock. They are less regular in form and of lower grade than the gravels within the confines of the canyons and extend but a short distance out into the Klehini Valley. Apparently the alluvial fans are devoid of notable placer concentrations, except possibly at their very heads.

MINING METHODS.

The strong relief and abundant water supply of the region afford splendid facilities for hydraulic mining. Ample hydraulic head can be developed with an unusually short length of flume, owing to the steep grades of the streams. The abundant local supply of timber makes the cost of flume construction low. It is therefore natural that equipment for hydraulic mining has replaced all the various devices that were tried out in the course of the earlier operations.

The chief obstacles to mining the creek gravels are presented by the great discharge of the streams during the mining season and the presence in the placers of boulders that are too large for sluicing. (See Pl. VI, *A* and *B*.) The bench deposits also contain large boulders, but there is no problem of water control connected with working them.

The streams are diverted from the placer ground by means of wooden flumes. The flume on Porcupine Creek is 24 to 30 feet wide and 6 feet deep and was originally over a mile long (Pl. VII, A). A section 2,000 feet long at the downstream end is now out of repair and not in use. The slope is about 5 inches in 100 feet. The water develops very high velocities in the flume during floods, so that the entire discharge of Porcupine Creek is generally handled successfully. At high stages a large amount of débris, including gravels and fairly large boulders, passes through the flume, which causes considerable wear and makes replanking necessary every year or two.

Similar diverting flumes were used in mining the placers of McKinley and Cahoon creeks. The available ground is now worked out. and the flumes are largely destroyed. (See Pl. VII, B.)

The diverting flume on Glacier Creek is only 8 feet wide and 5 feet high, but the grade is 6 to 10 inches in 12 feet, so that the discharge of the creek, about 1,000 second-feet at the maximum, is ordinarily handled successfully. (See Pl. VIII, A.) The flume is floored with 6-inch riffle blocks cut across the grain, which makes reflooring necessary only at long intervals. (See Pl. VIII, B.) The flume is 1,950 feet in length. It carries the water from the dam at the upper end of the placer ground to the head of an old channel on the alluvial fan east of the one normally followed by Glacier Creek.

Successful mining of the creek gravels depends primarily upon keeping the diverting flumes in full commission at all times. Temporary failure, even for a few hours, in time of flood may result in complete refilling of working pits with coarse wash. This has happened on both Glacier and Porcupine creeks in the past. The filling in the workings on Porcupine Creek during the flood of 1915 is estimated to have had an average depth of 12 feet for a distance of over half a mile. Much valuable time and labor is sacrificed in dead work in restoring workings to proper shape after each failure of the flumes. The operators are now paying special attention to maintenance of the flumes, and it is hoped that improved construction may eliminate this precarious feature entirely.

The large boulders in the placers, which are handled by crane hoists, are lifted in cable slings either whole or after being blasted; those small enough to be handled are thrown on tables which are hoisted and dumped mechanically. Hydroelectric power is used on Glacier Creek, steam power is used at the Porcupine plant, and power is delivered directly from a Pelton wheel to the winch on McKinley Creek.

The arrangement of sluice boxes and the methods of delivering the pay dirt to them and of disposing of tailings differ according to the situation of the placers and the general equipment of the plants. On Glacier Creek the gravels average between 45 and 50

A. DIVERTING FLUME, PORCUPINE CREEK.

B. WORKED-OUT SECTION OF MODERN McKINLEY CREEK CANYON.

Note position of abandoned canyon at head of waterfalls, same locality as shown in Plate V, *B.*

A. FLUME AND SPILLWAY, GLACIER CREEK DAM.

B. DETAIL OF GLACIER CREEK FLUME, SHOWING USE OF RIFFLE BLOCKS.

feet in depth. The upper gravels to a depth of 20 feet are piped off and stacked at one side by a 3-inch giant working under a hydraulic head of about 400 feet. The rest of the deposit is piped over a grizzly, which passes materials less than 1 foot in diameter to a hydraulic elevator. The elevator has a 12-inch throat and 20-inch tube and elevates the materials 45 feet to the sluice box. The box is 4 feet wide, 3 feet deep, and 128 feet long and is floored throughout with 60-pound steel rails set crosswise for riffles. Tailings are stacked with a 3-inch giant.

On Porcupine Creek the pay gravel is hydraulicked directly into sluice boxes set in·bedrock, and the tailings are carried through a flume set with block riffles for about 2,000 feet, where they are delivered to the creek bed at the lower end of the diverting flume.

The bench placers of McKinley Creek are about 200 feet above the level of the stream, so that there is abundant room for tailings. The face of the cut is about 60 feet high and 50 to 70 feet wide between bedrock walls. The gravels and glacial overburden are caved by giants, the heavy boulders are stacked behind cribbing at one side (Pl. VI, *B*), and all the finer material is hydraulicked directly into sluice boxes set in bedrock, which discharge over the rim of the modern canyon of McKinley Creek.

HISTORY OF DEVELOPMENT.

Gold was first discovered in the Porcupine district in 1898. The following year mining operations were begun on Porcupine Creek, and the great number of prospectors who assembled staked claims on other creeks of the district. Though gold-bearing gravels were found on Bear Creek, on several tributaries of the Salmon, and on the head of Takhin River, productive operations up to the present time have been almost entirely limited to Porcupine Creek and its tributaries. A little mining was done on Nugget Creek from 1902 to 1911. The ground was then abandoned, and no further work has been done. It is estimated by local operators that about $6,000 worth of gold was produced.

The gold production of Porcupine Creek and its tributaries from 1898 to 1903 is given in the following table by Wright:[1]

Gold produced in the Porcupine region, 1898–1903.

1898	$1,000	1902	$140,000
1899	9,000	1903	150,000
1900	50,000		
1901	110,000		460,000

[1] Wright, C. W., The Porcupine placer district, Alaska: U. S. Geol. Survey Bull. 256, p. 13, 1904.

The production from these creeks is said to have continued at the rate of about $150,000 a year until 1906, when the principal works were destroyed by an unusual flood. From 1907 to 1909 large operations were discontinued, and the only production was made by a few laymen, who worked small lots of ground by manual methods. Production on a large scale was resumed in 1910, and it is estimated that an average yearly production of $50,000 was maintained until 1915, when another disastrous flood occurred. The total output for the district from 1898 to 1916, inclusive, estimated on the basis of these very incomplete data, is about $1,200,000.

In 1908 the Porcupine Mining Co. was organized to exploit the placers of the main Porcupine on a large scale. It was financed in the Eastern States, and in view of the equipment installed in the next few years it must have had a large investment fund. The first move of the new company was to construct a flume a mile long, 24 feet wide, and 6 feet deep, supported on piles, to carry the waters of Porcupine Creek past the placer ground to be worked. This piece of construction, which required nearly a millon feet of lumber and several thousand piles, was completed late in the summer of 1909, and mining was begun at the lower end of the canyon. This company operated until August, 1915, when the lower part of the flume was demolished and the pits were filled in by a flood. During the last few years of operation the work was done under a receivership but with the same local management.

After the disastrous flood of 1915 the property and holdings of the company were taken over by the Alaska Corporation. This concern in 1916 repaired the upper section of the old flume, constructed a new high-line flume to deliver water to the giants, and began the reexcavation of the buried workings.

But little mining was done on the tributaries of Porcupine Creek before 1908, when the Cahoon Creek Gold Mining Co. began work on McKinley and Cahoon creeks. The operations of this company have continued to the present time, working out the placers on Cahoon Creek near its mouth and for about 2,000 feet down McKinley Creek below Cahoon. The plan followed in the operations on McKinley Creek was essentially similar to that employed on Porcupine Creek— the stream was diverted into a wooden flume, making the stream bed available for hydraulic mining. Below the workings on McKinley Creek the stream runs through a narrow box canyon. The difficulty of diverting the stream in this reach has thus far prevented the exploitation of the gravels of its bed. In 1916 the company was engaged in working the placers of a high glacially filled channel on the right side of McKinley Creek, opposite the reach of the modern channel that had been worked out.

A little mining was done on the head of Cahoon Creek in the early days of the camp, but these operations are said to have met with little success, and the hope of working this ground on any large scale has long since been abandoned. However, a little prospecting and " sniping " has been done from time to time by laymen.

The most notable development in the district in the last few years has been the investigation of the Glacier Creek placers by drilling and the installation of a large hydraulic plant to work this ground. The Glacier Creek placers were staked in the early days of the district, and repeated attempts to prospect and mine the ground by ordinary methods failed, owing to the depth of the gravels and the abundance of ground water. In 1911 the claims, which had been abandoned by the previous holders, were restaked and systematic drilling was begun. Drill sections across the valley bottom were made at close intervals for over a mile upstream from the margin of the Klehini Valley. On the basis of the results of this investigation extensive preparations were made to work the lower section of the valley, 4,200 feet long, by hydraulic methods. The installation of a very complete plant, including dams, flumes, pipe lines, giants, hydraulic elevator and other machinery, was finished in the midsummer of 1915, but operation was prevented for the rest of the season by the unusual floods of that year. In 1916 work was started early in the spring, but owing to damage done the workings by a flood in the later part of June the season netted but little productive operation.

SUMMARY OF MINING IN 1916.

Mining in the Porcupine district in 1916 was confined to Porcupine Creek and its main east tributary, McKinley Creek, and Glacier Creek. A single plant using hydraulic methods worked on each stream throughout the summer. The principal work done on Porcupine and Glacier creeks consisted of the installation of hydraulic equipment and the repairing of damage done to workings and equipment by floods, so that little productive mining was accomplished. Two small plants, using manual methods, worked on the lower Porcupine part of the summer. The plant on McKinley Creek opened up a new pit in an old elevated channel and worked out a considerable area, although the operations were hampered somewhat by the presence of an overburden of glacial till 60 feet thick, in which there are numerous large boulders.

About 50 men on an average were employed during the summer in the whole district, divided about equally among the three active creeks. Though the production for 1916 was small, it is reported that the plants had all been put in good repair and were ready for more extensive operations.

PROSPECTING IN ADJACENT DISTRICTS.

RAINY HOLLOW DISTRICT.

The Rainy Hollow district is in Canadian territory on the head-waters of Klehini River about 12 miles northwest of the mouth of Jarvis Creek.[1] About 40 lode claims are held on several more or less parallel ledges that trend in a northerly direction. The most valuable deposits have been formed by the replacement of limestone. The mineralization of the district is markedly different from that of the main Porcupine placer area and is more nearly allied with that of Bear Creek. The metals are chiefly copper, silver, zinc, and lead. Gold is present only in negligible amounts. The dominant gangue minerals are quartz, garnet, and tremolite.

REGION EAST OF CHILKAT VALLEY.

Local interest has attached for several years to prospecting for gold and copper in the mountain range between Chilkat and Chilkoot valleys, northwest of Haines. Fine specimens of gold-bearing quartz and bornite copper ore, said to represent large deposits in this range, are frequently brought into Haines. If the current reports of the size and richness of these deposits are correct this locality should quickly assume a high rank among those producing these metals.

GEOLOGY.

The salient geologic features of this group of mountains is the contact between the diorites of the Coast Range, which make up their main mass, and the group of altered volcanic rocks that form their western base and extend southeastward into Chilkat Peninsula and the Berners Bay region. Several bodies of schist and limestones are included in the diorite near its main contact with the bedded rocks. Considerable alteration is generally evident in the bedded rocks near the contact with the diorite. The limestones are thoroughly recrystallized, and other types of bedded rocks are converted into biotite schists. Quartz veins and aplite dikes are locally abundant near the diorite, and in places quartz veins extend far out into the invaded rocks.

VEINS AND MINERALIZATION.

The metamorphism of the volcanic rocks near Haines apparently comprises their complete recrystallization with partial segregation of iron in the form of magnetite. Farther from the intrusive con-

---

[1]Brewer, W. M., British Columbia Minister of Mines Ann. Rept. 1914, pp. 94–99, 1915.

tacts the volcanic rocks have largely preserved their original character but generally show more or less mineralization. Thin veinlets of bornite are very prevalent throughout the volcanic area. Quartz veins and reticulating quartz stringers carrying copper and iron sulphides occur in places. Some of the quartz veins contain gold, according to local reports.

The most promising gold and copper prospects are said to be located near the top of the range, about 10 miles from Haines. Both gold and copper ores purporting to come from extensive ledges at this locality were examined by the writer. The gold ore consisted of vein quartz which contained fine specks of native gold and a little pyrite. The copper specimens were chiefly bornite, broken from a vein apparently several inches across. The bornite is said to have a notable content of gold, as shown by assay.

The gold quartz is said to occur in a broad ledge along a diorite-limestone contact, and the copper ore in a vein several feet wide which cuts diagonally across it. The deposit occurs at an altitude of about 4,500 feet, in a small basin that is free from snow for only a short time each year.

In view of the widespread evidences of mineralization in this area it is a matter of great interest that prospectors are actively investigating it and that the results thus far achieved have stimulated hopeful expectations.

IRON-ORE DEPOSITS NEAR HAINES.

Rocks which are probably related in origin to both the volcanic group and the diorites of the Coast Range and which contain so much magnetite that they have been regarded as a possible source of iron ore crop out along the shore of Chilkoot Inlet and in the adjacent mountains near Haines. Their lithology has been well described by Knopf,[1] who studied them in 1908, as follows:

The rock mass exposed along the shore north of Haines is a remarkable occurrence geologically. Specimens collected from the finest-textured portions show a rock composed of a coarsely crystalline aggregate of feldspar, hornblende, and pyroxene, throughout which are scattered some visible grains of magnetite. The dark minerals (the hornblende and pyroxene) make up half the bulk of the rock. When examined microscopically the rock is found to consist of an allotriomorphic granular assemblage of plagioclase feldspar (bytownite), hornblende, and augite. Magnetite and apatite are present as accessory constituents in unusually large amounts. From this normal type of rock, which would be termed a gabbro, abrupt variations in texture and mineral composition are encountered. In places the cliffs for hundreds of feet are composed solidly of formless hornblende individuals 6 inches long by 3 inches

---

[1] Knopf, Adolph, The occurrence of iron ore near Haines: U. S. Geol. Survey Bull. 442, pp. 144–145, 1909.

broad. Commonly this hornblende rock contains more or less grayish-green augite admixed with it and is ramified by coarse white feldspathic dikelets or blotched by masses of gabbro. In places it even forms a breccia cemented by such material. Locally the hornblendite contains numerous lumps and particles of magnetite, which can easily be recognized by the characteristic bluish tarnish that they assume upon weathered surfaces. At no point along the shore, however, has the segregation of magnetite proceeded far enough to yield a solid body of iron ore, or even a body of ore of commercial grade.

The structure of these rocks shows that they are not of simple magmatic origin. The hornblende rock is everywhere traversed in many directions by light-colored feldspathic dikelets and by feldspathic zones with less distinct outlines. In the greater part of the area the dikes and feldspathic zones are closely spaced and by intersection divide the hornblende rock into a multitude of separate blocks. The dikes are plainly intrusive in the basic rock. The feldspathic zones likewise have been developed subsequent to its original consolidation, probably by magmatic solutions. The disposition of the younger rocks in the general mass seems patterned after a rather definite joint system. Clearly the hornblendic portions of the mass represent the original rock of the area and the dominantly feldspathic portions a later development related to the diorite intrusion.

The origin of the magnetite-bearing hornblende rocks is somewhat problematic. Knopf[1] holds that they "represent various modifications of a single intrusive mass of deep-seated origin." On the other hand, it seems possible that they have been produced by the recrystallization of basalts of the volcanic group that partly surrounds the area. Basaltic and coarse-textured hornblendic rocks alternate in the exposures along the shore northwest of Battery Point, and in places there is an apparent textural gradation from one type into the other. In this area the constant association of feldspathic dikelets with the hornblende rock is strikingly exemplified. Some of the feldspathic zones and dikes merge with the hornblende rock in a way that suggests contemporaneous crystallization of the ferromagnesian and feldspathic minerals. There is nothing to indicate that the ferromagnesian minerals of the hornblende rock differ in age from those of the feldspathic zones. Thus it might well be inferred that the texture and mineralogic character of the hornblende rocks were developed pyrogenetically at the time of the intrusion of the feldspathic dikes—presumably during some epoch in the general period of intrusion of the diorites of the Coast Range.

The economic possibilities of the deposits have been summed up by Knopf,[2] as follows:

---

[1] Op. cit., p. 144.   [2] Op. cit., p. 146.

The iron ore occurring near Haines consists of primary magnetite sparsely disseminated in a basic igneous rock composed of pyroxene and hornblende. An ore of this character would require fine crushing and concentration. A study of producing iron-ore properties where the adoption of such processes was necessary will therefore furnish valuable data for an analysis of the commercial possibilities of the Alaskan iron-ore deposits. Many factors enter into the problem but hardly need discussion at this time. At Lyon Mountain, N. Y., where crushing, drying, and electromagnetic concentration are necessary, the lowest-grade iron ore treated contains 34 per cent of metallic iron.[1] This is far above that of the highest-grade rock so far found at Haines. Geologically it is possible that richer bodies of magnetite may occur as segregations in the basic granular rocks that form the ridge extending northwestward from Haines. A magnetic survey of the area underlain by these rocks would undoubtedly prove a quicker and more economical way to test this possibility than the driving of expensive prospect tunnels.

---

[1] Eng. and Min. Jour., vol. 82, p. 916, 1906.

Other Publications

by

Miningbooks.com

- Placer Gold Deposits of Nevada
- Placer Gold Deposits of Utah
- Placer Gold Deposits of Arizona
- Gold Placers of California
- Browns Assaying
- Arizona Gold Placers and Placering
- Arizona Lode Gold Mines and Gold Mining
- Dredging for Gold in California
- Metallurgy
- Gold Deposits of Georgia
- Placer Examination: Principles and Practice
- Geology and Ore Deposits of the Creede District, Colorado
- Gold in Washington
- Placer Mining in Nevada
- Gold Placers and their Geologic Environment in Northwestern Park County, CO
- Placer Mining for Gold in California
- Geology and Ore Deposits of Shoshone County, Idaho
- Gold Districts of California
- Gold and Silver in Oregon
- The Porcupine Gold Placer District Alaska
- Gold Placer Deposits of the Pioneer District Montana
- Economic Geology of the Silverton Quadrangle, Colorado
- The Ore Deposits of New Mexico

- **Roasting of Gold and Silver Ores, and the Extraction of their Respective Metals without Quicksilver**

- **Geology and Ore Deposits of the Summitville District San Juan Mountains Colorado**

www.ingramcontent.com/pod-product-compliance
Lightning Source LLC
Chambersburg PA
CBHW032021190326
41520CB00007B/564